Contents:-

Page:-

1. Theorems of Circle–1-3
2. Theorems of quadrilaterals —-4-7
3. Problems of triangles —-- 9—14
4. Theorems of triangle —- 16—20
5. Hydrocarbons—21-23
6. Light —24—29

Prove that perpendicular drawn from the center of a circle intersect a chord, bisects the chord.

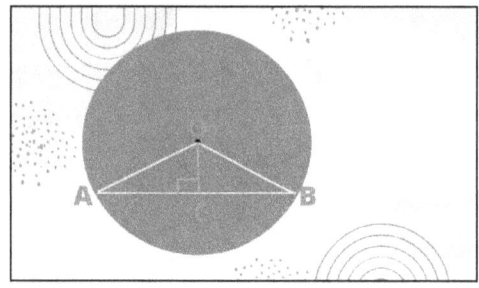

Answer:-

In the picture, a circle is drawn with center 'O'. AB is a chord of the circle. Perpendicular drawn from the center 'O' on the chord AB. i.e; OC⊥ AB

To prove that:- OC bisects AB, i.e; AC = BC

Construction:- OA and OB are drawn.

Proof:- In between Δs AOC and BOC we get,

∵ OA = OB [∵ radii of the same circle]

∵ OC = OC [∵ Common side of the Δs]

∵ ∠OCA = ∠OCB [∵ both are 90°]

∴ Δ AOC ≅ Δ BOC [By SAS congruence rule of the Δs]

∴ AC = BC [∵ By CPCT]   Proved

Prove that line joining the midpoint of a chord of a circle bisects the chord is perpendicular on the chord.

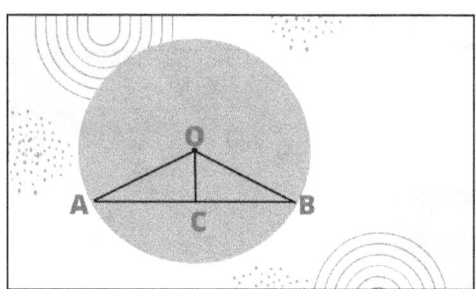

Answer:- In the picture, AB is a chord of a circle with center 'O' C is the midpoint of AB , OC is joined which bisects AB at point 'C' i.e; AC = BC …..(1)

To prove:- OC ⊥ AB

Construction:- OA and OB are joined.

Proof:- In between Δs AOC and BOC we get,

∵ OC = OC [∵ Common side of the Δs]

∵ AC = BC [∵ from (1)]

and OA = OB [∵ radii of the same circle]

∴ Δ AOC ≅ Δ BOC [∵ By SSS congruence rule of the Δs]

∴ ∠ OCA = <OCB [by CPCT] but they are adjacent angles.

∴ ∠OCA = <OCB = 180°/2 = 90°

∴ OC⊥ AB Proved.

Prove that, angle at the center of a circle is double the angle of the circumference standing on the same arc of the circle.

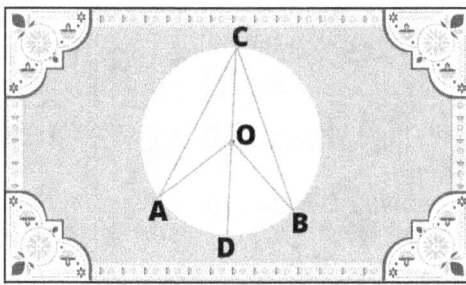

Answer:- In picture, <AOB and <ACB are the angles at the center and at the circumference respectively,standing on the same arc ADC of a circle with center 'O'

To prove that:- <AOB = 2<ACB

Construction:- OC is joined and extended upto D.

Proof:- ∵ In Δ AOC, we get

∵ OA = OC [∵radii of the same circle]

∴ ∠OCA = <OAC…..(1) [∵angles opposite to the equal sides are also equal]

Again,

In △ AOC

∵ <AOD = <OCA + <OAC [∵ sum of two remote interrior angles of a △ is equal to exterrior angle]

=> <AOD = <OCA+<OCA [using (1)]
=> <AOD = 2<OCA …..(2)

Similarly we can prove,

<BOD = 2<OCB ……..(3)

Now,

(2)+(3) we get,

<AOD+<BOD = 2(<OCA + <OCB)

=> <AOB = 2<ACB Proved.

Prove that equal chords of a circle are equidistant from the center.

Answer:-

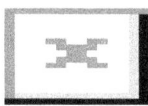

In the picture , AB and CD are two chords of a circle with center 'O' also AB=CD,and
OE⊥AB, OF⊥CD ,To prove:- OE=OF

Construction;- OA and OC are joined.

Proof:- ∵ OE⊥ AB
∴ AE = ½ AB [∵ ⊥ drawn from the center on the chord bisects the chord]

Similarly,

CF= ½ CD

∵ AB = CD [∵Given]
=>½ AB = ½ CD [∵dividing both sides by2
=>AE = CF

Now in between Δs OAE and OCF we get,

∵ AE = CF [∵ Proved]
∵ OA = OC [∵ radii of the same circle]
and, <OEA = <OFC [∵both are 90°]

∴ ΔOAE ≅ ΔOFC [∵by SAS rule of congruence of Δs]

∴ OE= OF [by CPCT]

Proved.

 Prove that equidistant chords of a circle are equal.

Answer:-

In the picture, AB and CD are two chords of a circle with center 'O' OE⊥AB, OF⊥CD, also OE = OF,
To prove:- AB=CD

Construction:- OA and OC are joined.

Proof:- ∵ OE⊥ AB
∴ AE = ½ AB [∵⊥ drawn from the center of a circle on a chord bisects the chord]

Similarly,

CF = ½ CD

Now in between Δs OEA and OFC we get,

∵ OE = OF [∵ given]
∵ ∠OEA =<OFC [∵ both are 90°]
and OA = OC [∵ radii of the same circle]
∴ Δ OEA≅ ΔOFC
∴ AE= CF [by CPCT]
=>½ AB =½ CD
=> AB = CD [∵ multiplying both sides by2]

Proved.

1. Prove that if one pair of opposite side of a quadrilateral is equal and parallel then it is a parallelogram.

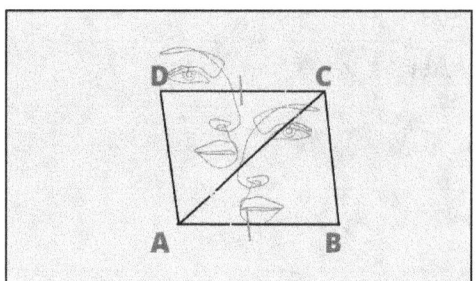

Answer:- Let, ABCD is a quadrilateral, whose AB ∥ CD ,also AB = CD, To prove that, ABCD is a parallelogram.

Construction:- Diagonal AC is joined.

Proof:- ∵ AB∥ CD and AC is the transversal

∴ ∠BAC = <ACD [∵ alternate angles are equal].....(1)

Now, in between Δs ABC and ACD we get,

∵ ∠BAC = <ACD [from….(1)]

∵ AB = CD [∵ given]

and, AC = AC [∵ common side of the Δs]

∴ Δ ABC ≅ Δ ACD [∵by SAS congruence rule]

∴ BC = AD   [∵ by CPCT]

and <ACB = <CAD

but , <ACB and < CAD are alternate angles and AC is the transversal between them,

∴ BC ∥ AD

Now, in quadrilateral ABCD, we find,

∵ BC ∥ AD, BC= AD [∵ Proved]
and AB∥ CA, AB= CD [∵ given]
So,

Opposite sides of quadrilateral ABCD are equal and parallel.

∴ ABCD is a parallelogram , Proved.

2. Prove that if opposite sides of a quadrilateral are equal, then it is a parallelogram.

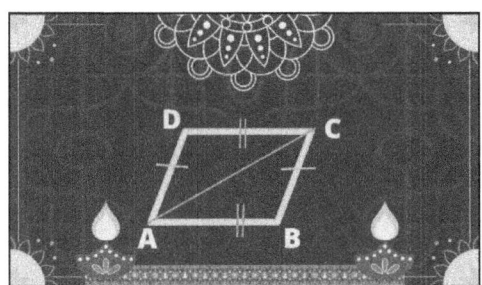

Answer:-
Let, ABCD is a quadrilateral,its
AB= CD and BC = AD To prove, ABCD is a parallelogram.

Construction:- Diagonal AC is joined.

Proof:-∵ In between Δs ABC and ADC we get,

∵ AB = CD [∵ given]

∵ BC = AD [∵ given]

and AC = AC [∵Common side of the Δs]

∴ Δ ABC ≅ Δ ADC[∵by SSS congruence rule]

∴ ∠BAC = < ADC[by CPCT]
,but they are alternate angles and AC is the transversal between
 them,

∴ AB∥CD, similarly we can prove

BC∥AD

Now in quadrilateral, ABCD we find

AB = CD   [∵given]
and BC = AD
and AB∥ CD   [∵ proved]
also BC∥ AD

So, opposite sides of the quadrilateral

ABCD are equal and parallel

∴ ABCS is a parallelogram, proved.

3. Prove that diagonals of a parallelogram bisect each-other.

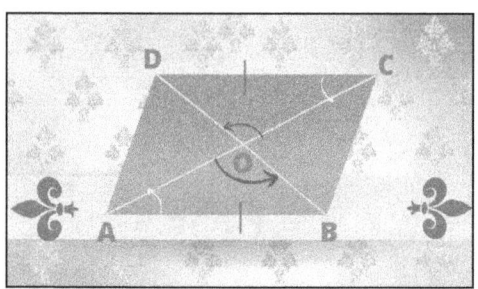

Answer:- Let, ABCD is a parallelogram, AC and BD are its diagonals intersect each other at point "O"

To prove, AO = OC
and BO = OD

Proof:- ∵ In between Δs AOB and COD
we get,

∵ AB = CD [∵ Opposite sides of the parallelogram]   ∵ ∠AOB = <COD [∵ vertically opposite angles]

and,
<OAB = <OCD [∵ alternate angles]

∴ Δ AOB ≅ ΔCOD [∵ by ASA congruence rule]
∴ AO = OC
and   [∵ by CPCT]
BO = OD
        Proved.

4. Prove that diagonals of a rectangle are equal.

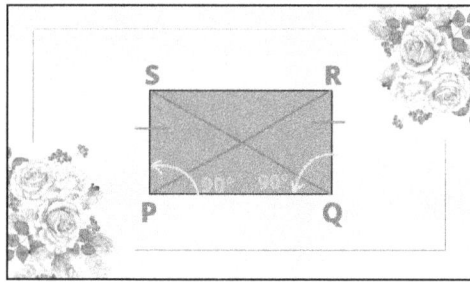

Answer:- Let, PQRS is a rectangle, PR and QS are its two diagonals.

To prove:- PR = QS

Proof:- ∵ In between Δs PQR and PQS
we get,

∵ PR = PS [∵ Opposite sides of the rectangle]

∵ PQ = PQ [∵ commonside of the Δs]

and

<PQR = <QPS [∵ABCD is rectangle, so both angles are 90°]

∴ Δ PQR ≅ ΔPQS

∴ PR = QS [by CPCT]
Proved.

5. Prove that if one angle of a rectangle is 90°, then all angles of the rectangle is right angle.

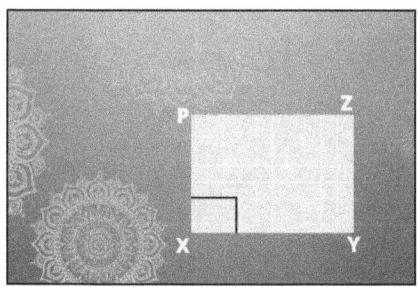

Answer:-
Let, XYZP is a rectangle, its <X=90°

To prove;-<X=<Y=<Z=<P=90°= right angle

Proof:- ∵ ABCD is a rectangle,

∴∠X +∠Y=180°[∵sum of adjacent angles of a rectangle is 180°]

=> 90° + <Y = 180°[<X=90°given]

=> <Y = 180°- 90°

=> <Y= 90°

Similarly,
we can prove, <Z =<P=90°

Now, in rectangle XYZP

<X=<Y=<Z=<P=90°= right angle ,proved

Problems of exterrior angle of Triangles:-

1. From the picture, Find <ACD=? If <A= 50° and <B=60°

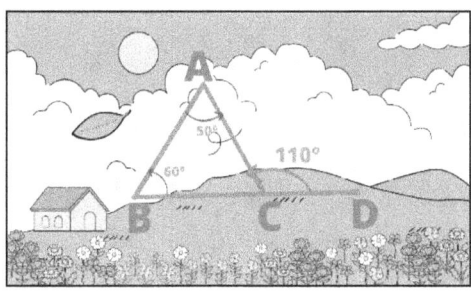

Answer:-
Here,

ABC is a triangle
∴ ∠ACD = <A + <B [∵sum of two remote interrior angles of a Δ is always equal to its exterrior angle]
= 50°+ 60°
= 110°

2. From the picture, Find <BAD, if <B = 60°
<C=70°

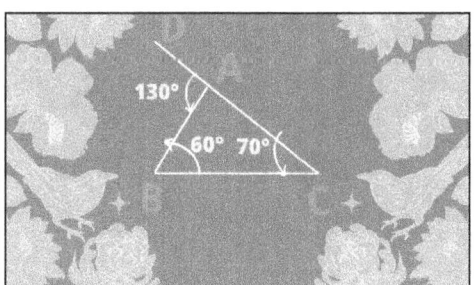

Answer:-

Here,

ABC is a triangle

∴ ∠BAD = <B+<C [∵ sum of two remote interior angles of a Δ is equal to exterrior angle]

= 60° + 70°

= 130°

3. In the picture, <Z=100° <Y= 40° then find <X=?

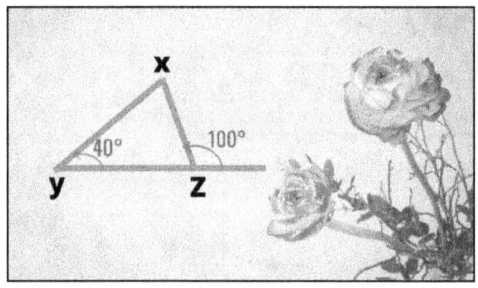

In the picture,

XYZ is a Δ ,whose <Z = 100°,<Y= 40°

To find:- <X = ?

Solution:-

∴ ∠X+<Y=<Z [∵ sum of two remote interrior angles is equal to exterrior angle]
=> <X + 40° = 100°
=> <X = 100° - 40°
=> <X = 60°

4. From the figure, find x=? and y = ?

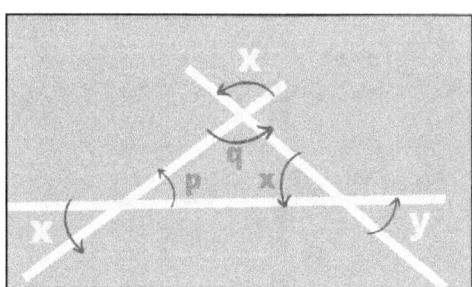

Answer:-

From the figure, we will find the values of x=? and y = ?

∵ x = y [∵ vertically opposite angles]...(1)
∵ x = p [∵ vertically opposite angles]...(2)
also q = x [∵ vertically opposite angles]...(3)

Now,
 x + p + q = 180° [∵ sum of three angles of a △ is 180°]

=> x + x + x = 180° [usinf (1),(2), and (3)]

=> 3x = 180°
=> x = 180°/3
=> x = 60°

Again,
from (1)
 y = x
 => y = 60° [putting the value of x]

∴ Requird x = 60°
 y = 60°

Area problems of triangles:-

1. If height of a triangle is 4 cm, and it's base is 7cm, find it's area.

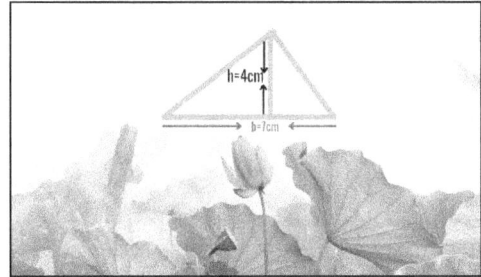

Answer:- Here,

 height of triangle(h) = 4cm

 base(b) = 7 cm

∴ Area of the triangle(A) = ½ b.h

$$= ½.4.7$$

$$= 2×7 \text{ cm}^2$$

$$= 14 \text{ cm}^2$$

2. The area of a triangle is 40 m² and its height 5m then find its base.

Answer:- Here,

Area of the triangle(A) = 40 m²

height of the triangle(h)= 5 m

∴ base of the triangle(b) = 2×A /h

$$=(2×40)/5 \text{ m}$$

$$= 2×8 \text{ m}$$

$$= 16 \text{ m}.$$

3. Area of a triangle is 100 m² its base 8 m then, find its height.

Answer:- Here,

Area of the triangle (A) = 100 m²

base of the triangle(b) = 8 cm

∴ Height of the triangle(h) =(2×A)/b

$$= 2×100/8$$

$$= 200/8$$

$$= 25 \text{ m}$$

4. Find the area of an equilateral triangle, whose each side is 8 cm.

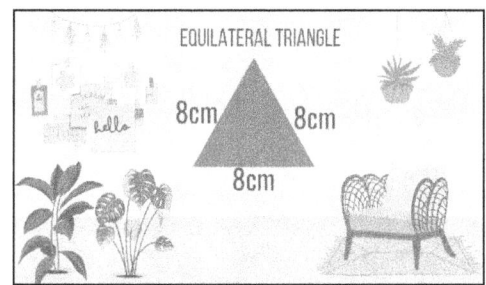

Answer:- Each side of the equilateral Δ(a)

$= 8$ cm

∴ Area of the equilateral Δ $= \sqrt{3}/4 \, a^2$

$= (\sqrt{3} \times 8 \times 8)/4$

$= \sqrt{3} \times 8 \times 2 \text{ cm}^2$

$= 16\sqrt{3} \text{ cm}^2$

5. The measures of each equal sides of an isoscles is 4 m and its base is 6 m, find the area of the triangle.

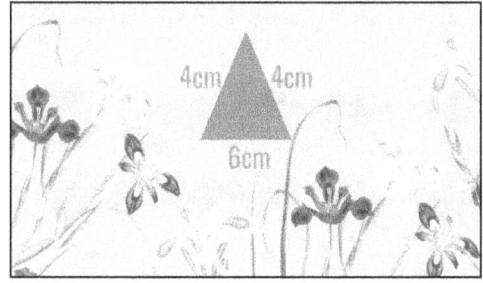

Answer:- Here,

∵Isoscles Δ has two equal sides.

∴1st side of isosceles Δ (a)= 4m

∴ 2nd side of isosceles Δ (b) =4m

and 3rd side of the Δ(c) = 6 m

Now,

Semi perimeter/half perimeter of the Δ(s)

= (a+b+c)/2

= (4+4+6)/2

= 14/2

= 7 m

∴ Area of the of the isosceles Δ is

= √ s.(s-a)(s-b)(s-c

= √ 7.(7-3).(7-3)(7-6)

= √ 7.4.4.1

= 4√7 m²

6. The three different sides of a scalene triangle is 3cm, 4cm and 5cm find its area.

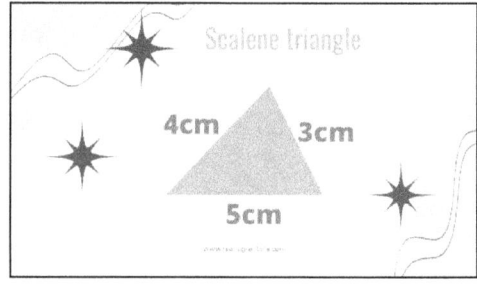

Answer :-

Here,
    1st side (a) = 3 cm
    2nd side (b) = 4 cm
and 3rd side (c) = 5 cm

∴ half perimeter of the scalene Δ
   (s) = (a+b+c)/2
    = (3+4+5)/2
    = 12/2
    = 6 cm

Now area of the scalene Δ
   = √ s.(s-a).(s-b).(s-c)

   = √ 6.(6-3).(6-4).(6-5)

= √6.3.2.1

= √6.6

= 6 cm²

1. Prove that sum of thee angles of a triangle is 180°

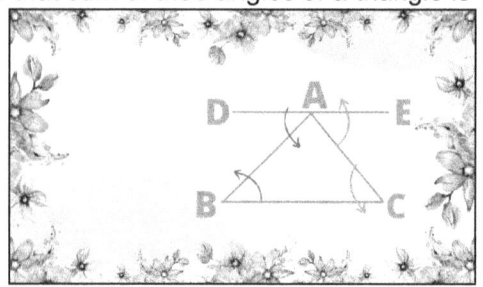

Answer:-

3. In the picture,

ABC is a △ , ∠A , ∠B and ∠C are its three angles,

4. To prove:-
5. ∠A + ∠B + ∠C = 180

6. Construction:- Through point "A" DE∥ BC is drawn.

7. Proof:- ∵ In △ ABC,

8. DE ∥ BC and AB is the transversal,

9. ∴ ∠B = ∠DAB [∵ alternate angles are equal].......(1)
10. Again,

11. DE ∥ BC and AC is the transversal,

12. ∴ ∠C = ∠EAC [∵ alternate angles are equal].......(2)

13. Now (1)+(2) we get,

14. ∠B + ∠C = ∠DAB + ∠EAC

15. again, adding ∠BAC in both sides of the above equation we get,

16. ∠BAC + ∠B + ∠C = ∠DAB + ∠BAC + ∠EAC

17. => ∠BAC + ∠B + ∠C = ∠DAE
18. => ∠BAC + ∠B + ∠C = 180° [∠DAE is a straight angle, so it measures 180°]

19. ∴ ∠A + ∠B + ∠C = 180° Proved

2.Prove that sum of two remote interior angles of a triangle is equal toi its exterrior angle.

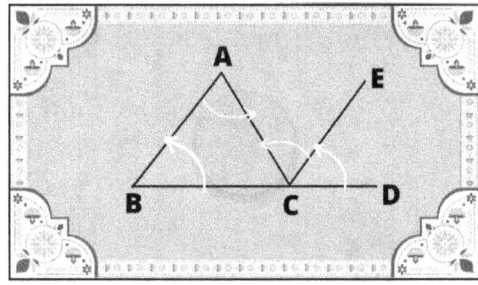

Answer:-

20. In the picture,

ABC is a Δ, ∠A and ∠B are its two remote interior angles,and ∠ACD is its exterrior angle.

21. To, prove :- <A+<B = < ACD

22. Construction:- Through " C" BA∥CE   is drawn.

Proof:- In ΔABC

∵ BA∥ CE and AC is the traversal,

23.   ∴ ∠A = <ACE [∵ alternate angles are [equal].....(1)

24. again,

25.   BA∥ CE and BD is the transversal,

26.   ∴ ∠ B = <ECD [∵ corresponding angles are equal].....(2)

27. Now, (1)+(2) we get,

28.   <A + <B = < <ACE + <ECD
29.   => <A + <B = < ACD
                     Proved.

3. Prove that sum of all the angles of a quadrilateral is 360°

Answer:-

In the picture,

ABCD is a quadrilateral,

30. To, prove that <A + <B + <C +<D =360°

31. Construction:- Diagonal AC is joined.

32. Proof:- ∵ In Δ ACD, we get,

33. ∵ ∠ACD + ∠ADC + ∠CAD =180° [∵ sum of three angles of a Δ is 360°]....(1)

34. Again,

35. In Δ ABC , we get,

36. ∵ ∠ABC +<ACB + <BAC =180° [∵ sum of three angles of a Δ is 180°].....(2)

37. Now,
38. (1)+(2) we get,

39. ∵ <CAD + <BAC)+<ABC + (<ACD+   <ACB)+ <ADC = 360°

40. ∴ < A + < B + < C + < D =360° Proved.

4. Prove that opposite sides of a parallelogram are equal.

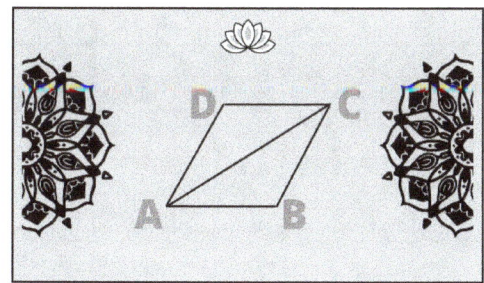

Answer:-

In the picture,

ABCD is a parallelogram, It has four sides AB, BC, CD and AD

- To, prove that,
- AB= CD and BC = AD
- Construction :-Diagonal AC is joined.
- Proof:- ∵ In between Δs ABC and ADC
- we get,
- ∵ ∠BAC = ∠ACD [∵ AB∥CD and AC is the transversal, so alternate angles are equal]
- ∵ ∠ACB = ∠CAD [BC∥ AD and AC is the transversal, so alternate angles are equal]
- Again,
  AC = AC [Common side of the Δs]
- ∴ Δ ABC = Δ ADC
- ∴ AB = CD   [CPCT]
- and BC = AD
                Proved.

5. Prove that opposite angles of a parallelogram are equal.

Answer:- In the picture,

PQRS is a parallelogram.
<P, <Q, <R and <S are its four angles.

To prove that, <P = <R
and <Q = <S

Proof:- ∵ PQRS is a parallelogram,

∴ ∠P + ∠Q = 180° [sum of adjacent angles of a parallelogram is 180°]
............(1)

similarly,

<Q + <R = 180° [∵ sum of adjacent angles of a parallelogram is 180°]
.............(2)

Now from equations (1) and (2) we get,

<P + <Q = <Q + <R
=> <P = <R [∵ subtracting ∠Q from both sides]

similarly we can prove that,

<Q = <S

Proved

What are hydrocarbons?

Answer:- The binary compounds which are formed when carbon reacts with hydrogen is called hydrocarbons. e.g; Methane(CH_4), Ethane(C_2H_6), Ethene(C_2H_4), Propyne(C_3H_4))

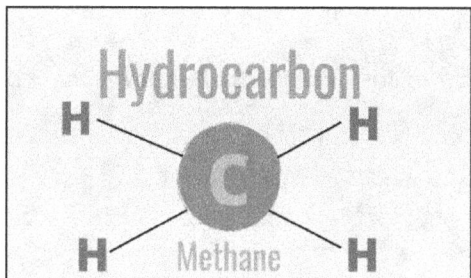

How Hydrocarbons are classified?

Answer:- Hydrocarbons are mainly divided in two types (1) Aliphatic or acyclic. (2) Alicyclic. They are further divided as per following picture shown:-

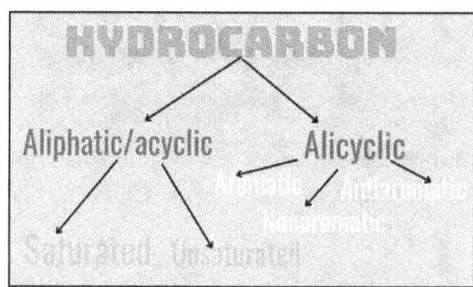

What are aliphatic or acyclic hydrocarbons?

Answer:- The hydrocarbons which have simple long chain structure is known as aliphatic or acyclic hydrocarbons. e,g; Butane(C_4H_{10}), Pentane(C_5H_{12})

What are alicyclic hydrocarbons?

Answer:- The hydrocarbons which are cyclic in order may have carbon ring or not in their structure are known as alicyclic hydrocarbons. e,g; Benzene(C_6H_6), Cyclo pentane(C_5H_{12})

What do you mean by saturated and unsaturated hydrocarbons?

Answer:- The hydrocarbons which have only single bonds in their structure are generally stable are known as saturated hydrocarbons or alkanes. e,g; Methane(CH_4), Ethane(C_2H_6). The general structural formula of alkanes is C_nH_{2n+2} where n= 1,2,3,4,.......

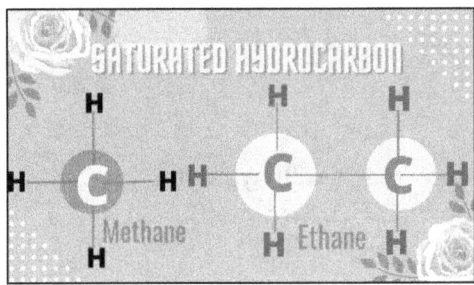

The hydrocarbons which have double or triple bonds in their structure is known as unsaturated hydrocarbons. e,g; Ethene(C_2H_4), Ethyne(C_2H_2)

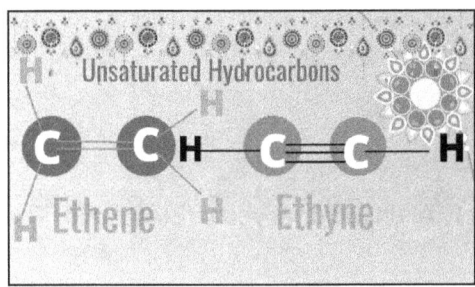

What are alkenes and alkynes ?

Answer:- The hydrocarbons which contain only double bonds in their structure is known as alkenes. e.g; Ethene(C_2H_4) Propene(C_3H_6). The general formula of alkenes is (C_nH_{2n}), Where n = 2,3,4……
and n≠1

The hydrocarbons which have only triple bonds in their structure is known as alkynes. e,g; Ethyne(C_2H_2), Propyne(C_3H_4)
The general formula of alkynes is (C_nH_{2n-2})
where, n=2,3,4…… and n≠ 1

How can we write the structural formulae of alkanes family?

Answer:- The general formula of alkanes is C_nH_{2n+2}

If we take n=1 then, C=n =1
and H=2n+2 will be H=2×1+2 =4
Then, the formula of the alkane will be CH_4 which is known as methane.
Similarly,
If we take, n=2 then, C=n=2
and H=2n+2 will be H=2×2+2 =4+2=6

So, alkane will be C_2H_6 which is known as ethane, in this way we can find more alkanes.

How can we write structual formulae of alkenes family?

Answer:- The general formula of alkenes is C_nH_{2n}

If n=2 then C = n =2
and H=2n =2×2=4
So, alkene will be C_2H_4 which is known as ethene.
Again if n=3 then, C= n =3
and H=2n =2×3=6
So, alkene will be C_3H_6 which is called propene.

How can we write structual formulae of alkynes

Answer:- The general formula of alkynes is
C_nH_{2n-2}
If we n=2 then, C=n=2
also H=2n-2 =2×2 -2 =4-2=2

So, the alkyne will be, C_2H_2 which is known as ethyne.

Again, if n=3 then C=n=3
also H=2×3-2=6-2=4
So, the alkyne will be C_3H_4 which is known as propyne.

What are parafins and olefins?

Answer:- Alkanes contain only single bonds in their structure so they are very stable and can not be broken, so they are called parafins. e.g; CH_4 , C_2H_6 etc.

Alkenes contain only double bond in their structure so they sre unstable and can be broken to alkanes under specific conditions, also oil can be obtained from them, so they are called olifins. e.g; C_2H_4 , C_3H_6 etc.

What is light?

Answer:- Light is a form of energy which give us sensation to look at.

What are photons?

Answer:- The particles of light are known as photons.

what is medium?

Answer:- Medium is a substance or a path through which any energy like light or sound or heat can pass easily without any obstacle.

What are the different types of medium through which light can pass?

Answer:- Generally light can pass in two different mediums and they are:-

(1) Rarer medium
(2) Denser medium.

What are rarer and denser medium?

Answer:- The medium through which light rays pass more faster is known as rarer medium. e.g; Air.

The mediums which are heavy and light rays pass a little slower than rarer medium are known as denser medium. e.g; Water, Glass.

What is reflection of light?

Answer:- Light ray which is travelling from a source through a medium (say air) in a straight line if is obstructed by another medium or a surface then it bounce back from that medium or surface to the its 1st medium, this property of light is called reflection of light.

What are the laws of reflection?

Answer:- There are two laws of reflections:-

(1) Incident ray,reflected ray, normal,and point of incidence all lie in same plain.

(2) The angle of incidence and angle of reflection are always equal.i.e; <i = <r

What is irregular reflection?

Answer:- When light rays falls on a rough surface it scattered and bounce back from several points not a single point, this type of reflection is called irregular reflection.

What is refraction of light?

Answer:- Light ray which is coming from a source through a medium in a straight line when it enters in another medium,then it slightly change the direction of its path on the contact point of two medium, this property of light is called refraction of light.

What are the laws of refraction?

Answer:-
There are two laws of refraction:-
(1) Incident ray, refracted ray, normal and point of incidence all lie in same plain.

(2) The ratio of the sin of incidence angle to the sin of refraction angle is always a constant, and this constant is called refractive index of 2nd mediium with respect to the 1st medium, and is denoted by $^1\mu_2$. This law is also known as Snell's law.

Describe reflection of light with ray diagram .

Answer:-

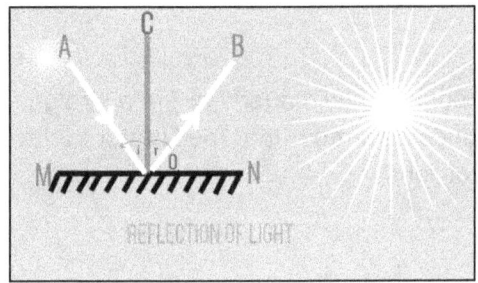

In the picture,

MN = Plain mirror.

AO = Incident ray.

OB = Reflected ray.

OC = Normal.

O = Point of incidence.

<AOC = <i = Angle of incidence.

< BOC = <r = Angle of reflection.

According to the laws of reflection,

<i = <r

and AO, OB, OC and O all lie in same plain.

Describe refraction of light with ray diagram.

Answer:-

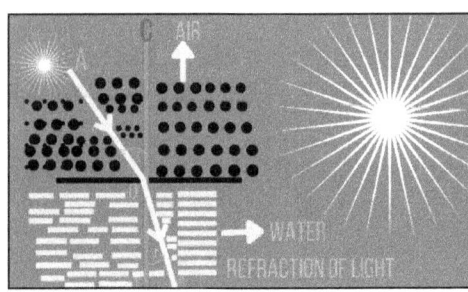

In the picture,

AO = Incident ray.

OB = Refracted ray.

<AOC = Angle of incidence.

<BOD = Angle of refraction.

CD = Normal.

O = Point of incidence.

According to the law of refraction,

AO, OB, CD and O all lie in same plain.

also

Sini /Sinr = $^1\mu_2$, where $^1\mu_2$ is a constant, known as refractive index of second medium with respect to first medium. This is also called Snell's Law.

What is real image?

Answer:- Light rays coming from a point source after reflection or refraction if they actually meet a fixed point then that point is called the real image of the first point. Real image has existence and can be catch in a screen.

What is virtual image?

Answer:- Light rays coming from a source after reflection or refraction if they appear to meet from a fixed point, then that point is called the virtual image of the first point.
Virtual image has no existence and can not be catch in a screen.

What is spherical mirror?

Answer:- Spherical mirror is a part of globe which is made of glass if we imagine.

How spherical mirror is classified?

Answer:- Spherical mirror is classified in two different catagories:-

(1) Concave mirror.
(2) Convex mirror.

Define Concave mirror.

Answer:- A spherical mirror whose lower plain surface acts as a reflecting surface is known as Concave mirror.

In the picture, MN is a concave mirror.

Define Convex mirror.

Answer:- A spherical mirror whose upper plain surface acts as a reflecting surface is known as Convex mirror.

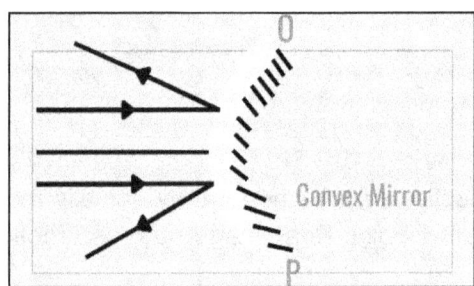

In the picture, OP is a convex mirror.

www.ingramcontent.com/pod-product-compliance
Lightning Source LLC
Chambersburg PA
CBHW080439220526
45465CB00009B/3355